Gettysburg is the scene of the decisive battle, a turning point of the American Civil War.

W9-DAQ-889

Situation, Spring 1863 7

The situation in which the Confederacy found itself called for decisive action. Confederate General Robert E. Lee decided to carry the war into northern territory.

Organization of the Army . . 8
Weapons and Tactics 78

Plan of Campaign 13
The First Day - July 1 18
The Second Day - July 2 31
The Third Day - July 3 50
End of Invasion 76

Lincoln and Gettysburg . . . 82

The aftermath of battle was a trying time for the residents of Gettysburg. A few months after the battle, Abraham Lincoln delivered his famous Gettysburg Address at the dedication of the national cemetery set apart as a burial ground for the soldiers who died in the conflict.

Gettysburg Cyclorama 92
The Battlefield Today 93

Gettysburg National Military Park has over 1,400 monuments and hundreds of cannon along over 40 miles of scenic roads. The Visitor Center has a 22,000 square foot museum, films and the restored Gettysburg Cyclorama.

Text from the Official Gettysburg National Park Handbook by Frederick Tilberg, 1954 and 1992. Additional text and captions by Scott Hartwig, Supervisory Historian, and John Heiser, Historian, Gettysburg National Military Park. Color cyclorama photos by Bill Dowling. Designed and edited by Henry Hird III ©2018 Historic Print & Map Company Printed in the USA

Suggested readings, additional credits and web links at **www.gettysburghandbook.com** ISBN 978-0-9729463-8-4

Gettysburg

ON THE GENTLY ROLLING FARM LANDS surrounding the little town of Gettysburg, Pennsylvania, was fought one of the great decisive battles of American history. For 3 days, from July 1 to 3, 1863, a gigantic struggle between 75,000 Confederates and 93,000 Union troops raged about the town and left 51,000 casualties in its wake, Heroic deeds were numerous on both sides, climaxed by the famed Confederate assault on July 3 which has become known throughout the world as Pickett's Charge. The Union Victory gained on these fields ended the last Confederate invasion of the North and marked the beginning of a gradual decline in Southern military power.

Here also, in November 1863, President Abraham Lincoln delivered his famous Gettysburg Address at the dedication of the Soldiers' National Cemetery.

3

THE BATTLE O

From the Original Picture Painted for the State of Penn

GETTYSBURG.

under award of Commission appointed by the Legislature.

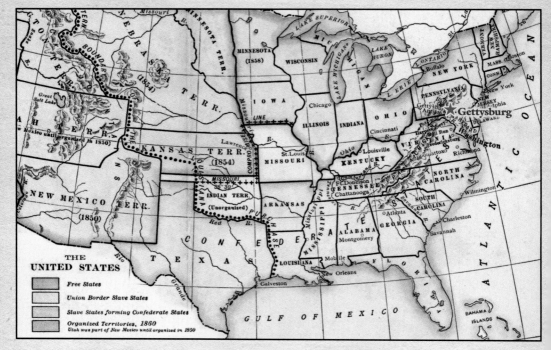

THE UNITED STATES

Free States

Union Border Slave States

Slave States forming Confederate States

Organized Territories, 1860
Utah was part of New Mexico until organized in 1850

1860

Nov 6 Abraham Lincoln wins election for President

1861

Feb 9 Jefferson Davis elected President of Confederate States of America

Mar 4 Abraham Lincoln inaugurated as President of the United States

Apr 12-13 Fort Sumter bombarded. Major Robert Anderson surrenders, no casualties.

July 21 First Battle of Bull Run, Manassas, Virginia - Confederate victory

Aug 10 Battle of Wilson's Creek, Mo - Confederate victory

1862

Feb 16 Ft. Donelson, Tn captured by Union Gen. U.S. Grant

Mar 7-8 Battle of Pea Ridge, Arkansas - Union victory

Mar 9 Ironclad U.S.S. Monitor vs. C.S.S. Virginia (Merrimack).

Apr 6-7 Battle of Shiloh, Pittsburg Landing, Tn - Union victory

Apr 10-11 Fort Pulaski, Savannah, Ga, is captured by Union

Apr 24-25 Farragut's U.S. Navy ships capture New Orleans

May 31 - June 1 Battle of Seven Pines, Va - Confed. Gen. Robert E. Lee becomes commander of Army of Northern Virginia

June 27 Battle of Gaines' Mill, Va, Confederate victory

July 1 Battle of Malvern Hill - Union victory.

Aug 29 Second Battle of Bull Run, Manassas, Va - Confederate victory,

Sept 14 Battle of South Mountain, Md - Union victory

Sept 17 Battle of Antietam in Md Lee's first northern invasion, no clear winner, 25,000 casualties.

Sept 22 Lincoln's Preliminary Emancipation Proclamation.

Oct 3-4 Battle of Corinth, Ms - Union victory

Dec 13 Battle of Fredericksburg, Va - Confederate victory

Dec 31- Jan 2 Battle of Stones River, Tenn. - Union victory

1863

May 3-4 Battle of Chancellorsville, Va. - Confed. victory that emboldens Gen. Lee to invade the North again.

May 16 Battle of Champion Hill, Miss. - Union victory

May 18 - July 4 Union siege of Vicksburg, Miss.

May 23 - July 9 Siege of Port Hudson, Louisiana

June 9 Battle of Brandy Station, Va - largest cavalry clash

July 1-3 **BATTLE OF GETTYSBURG** General Robt E. Lee is defeated by Union Gen. George Meade that forces Lee's retreat into Virginia

July 4 Surrender of Vicksburg to General U. S. Grant

July 9 Surrender of Port Hudson to Union Gen. Banks, closes the Mississippi River to the Confederacy

THE CIVIL WAR *continued another two years, ending with the Confederate surrender at Appomattox Court House in Virginia on April 9, 1865.*

Flag of the
United States in
1863.

President
Abraham Lincoln

National Flag
of the Confederate
States in 1863.

President
Jefferson Davis

The Situation, Spring 1863

The situation in which the Confederacy found itself in the late spring of 1863 called for decisive action. The Union and Confederate armies had faced each other on the Rappahannock River, near Fredericksburg, Va., for 6 months, The Confederate Army of Northern Virginia, commanded by Gen. R. E. Lee, had defeated the Union forces at Fredericksburg in December 1862 and again at Chancellorsville in May 1863, but the nature of the ground gave Lee little opportunity to follow up his advantage. When he began moving his army westward, on June 3, he hoped, at least, to draw his opponent away from the river to a more advantageous battleground. At most, he might carry the war into northern territory, where supplies could be taken from the enemy and a victory could be fully exploited. Even a fairly narrow margin of victory might enable Lee to capture one or more key cities and perhaps increase northern demands for a negotiated peace.

Confederate strategists had considered sending aid from Lee's army to Vicksburg, which Grant was then besieging, or dispatching help to General Bragg for his campaign against Rosecrans in Tennessee. They concluded, however, that Vicksburg could hold out until climatic conditions would force Grant to withdraw, and they reasoned that the eastern campaign was more important than that of Tennessee.

Both Union and Confederate governments had bitter opponents at home. Southern generals, reading in Northern newspapers the clamors for peace, had reason to believe that their foe's morale was fast weakening. They felt that the Army of Northern Virginia would continue to demonstrate its superiority over the Union Army of the Potomac and that the relief from constant campaigning on their own soil would have a happy effect on Southern spirit. Events were to prove, however, that the chief result of the intense alarm created by the invasion was to rally the populace to better support of the Union government.

INFANTRY

CAVALRY

ARTILLERY

Soldiers on Foot

Soldiers on Horseback

Soldiers with Cannons

Organization of the Armies

To the non-military buff, the organization and terminology used for Civil War armies can be very confusing. The "Army of the Potomac" was the main Union Army in the eastern theater of the war and the "Army of Northern Virginia" was the main Confederate Army. Both of the armies that fought the Battle of Gettysburg were organized in a similar fashion including the structure of corps, divisions, and brigades. Because the war had to be fought over a large area of the South, the Union and the Confederacy both had several armies, each fighting in different "departments" or sections of the country. But what were these different organizations and how did they all fit in to one huge force?

The Federal government and the Confederate government both had war departments, which oversaw the organization, supply, and movements of their respective armies. Civil War-era armies were organized according to military manuals adopted by the Federal War Department prior to 1861. Each army was a structured organization that included a general headquarters, infantry, artillery, cavalry, signalmen, engineers, quartermaster and commissary departments. The largest single organization of an army was a corps (pronounced "core"). The Union Army at Gettysburg had seven infantry corps and a cavalry corps, each commanded by a major general. The Confederate Army had three infantry corps, each commanded by a lieutenant general. Typically, a Confederate corps was much larger than a Union corps. A corps included three infantry divisions with an artillery brigade in the Union army while the Confederate army called it an artillery battalion. The Army of the Potomac had distinguishing symbols called corps badges to identify one corps from another. The badges were actually small cloth cut-outs shaped like crosses, spheres, stars, and quarter moons, and made in three different colors- red, white, and blue, each color specific to a division of the corps. Confederates had no corps badges or particular symbols for their organizations.

Union army corps badges

ARMY OF THE POTOMAC	ARMY OF NORTHERN VIRGINIA
93,000 soldiers at Gettysburg	75,000 soldiers at Gettysburg
divided into 8 Corps	divided into 3 Corps
(7 infantry corps & 1 cavalry corps)	(3 infantry corps & 1 cavalry division)
1 Corps = approximately 10,000 men	**1 Corps = approximately 22,000 men**
each Corps divided into 3 Divisions	each Corps divided into 3 Divisions
1 Division = 3,000 to 5,000 men	**1 Division = 6,000 to 8,000 men**
each Division divided into 2 to 3 Brigades	each Division divided into 4 Brigades
1 Brigade = 1,500 officers and men	**1 Brigade = 1,600 officers and men**
each Brigade divided into 4 to 6 Regiments	each Brigade divided into 4 to 6 Regiments
1 Regiment = average 307 men	**1 Regiment = average 330 men**
each Regiment divided into 10 Companies	each Regiment divided into 10 Companies
1 Company = average 30 men	**1 Company = average 33 men**

The infantry division was commanded by a major general and composed of two to four infantry brigades. The brigade, commanded by a brigadier general or colonel, was composed of four to six regiments, and was the primary tactical organization on the battlefield. The Confederate War Department made attempts to have brigades composed of regiments from one state, organized by state affiliation, such as General Joseph Kershaw's brigade composed of all South Carolina regiments. The Union Army did not always make such conscious choices, though there were some brigades which acquired interesting nick names due to their ethnic origin or locality from which they hailed.

For the infantryman or cavalryman, the regiment was the most important unit. Led by a colonel, lieutenant colonel and major, a full strength regiment numbered over 1,000 officers and men. Attrition due to disease and battle losses meant considerably lower personnel in each regiment by the time of the Battle of Gettysburg, where some regiments barely mustered two-hundred men. A regiment was divided into ten companies of 100 men each at full strength. One company was divided in half as two platoons. One company was led by a captain with two lieutenants who each commanded a platoon. Platoons were divided into squads, led by a sergeant or corporal. Regiments fought in a "battle line" or in some cases a "skirmish line", which was a general open rank tactic used to feel out the strength of an enemy force.

These are illustrations of uniforms they way they were supposed to be. In reality there was a large array of uniform styles and colors for both the Union and the Confederacy.

At the outbreak of the Civil War, there was a standing force of "regular" units in the United States Army. State militias were called into service, but there was a need to Federalize these units so that they could get pay from the United States government and serve outside of state borders. Each state was given a quota of "volunteer regiments" to be raised for service lasting from three months to three years. The South faced a similar dilemma. Southern states raised and supplied the Confederate armies with volunteer regiments. By 1863, many of the regiments in both armies had been in service since 1861 and were still composed of mostly volunteer soldiers, though the first "conscripts" or men required by state law to serve in the military defense of a state, had begun to appear in Southern units. A regiment's flag, or "regimental colors", were painted with the regiment's number and state affiliation, usually followed by "VOLUNTEER INFANTRY". The term volunteer was a symbol of pride for soldiers on both sides.

Confederate Battle Flag was carried by regiments of infantry, cavalry, and artillery

The most widely used manual for small units (regiments) was *Rifle and Light Infantry Tactics*, written by William J. Hardee. The manual specified the proper placement of officers, the rank and file, the manual of arms, basic marching orders, and other requirements. His manual was re-written for Confederate use in 1861 when Hardee resigned his commission from the United States Army and joined the Confederacy. Other manuals of organization and drill were used, but "Hardee's Tactics" continued to be the most popular and widely used manual throughout the war.

The artillery was usually organized by regiments as well, except that each company was called a battery. A battery consisted of over 100 soldiers, armed with six cannon per battery. Confederate batteries were smaller, some having only four cannon. Batteries were assigned independently from their regiments to specific artillery brigades (Union) or battalions (Confederate) or to the artillery reserve of an army. Both of the armies had an artillery reserve which was an organization of extra batteries to be placed where needed. The Union army had one large artillery reserve force. The Confederate army had one reserve group per corps, but the number of guns was still smaller than the number of Union cannon.

The Union army had flags for the different parts of the army. Shown here is the flag used by the Army of the Potomac, an infantry regiment flag, an artillery regiment flag and a cavalry regiment flag.

A cavalry regiment was organized in a similar fashion to the infantry and artillery. Ten to twelve companies or "troops", made up one regiment. The regiment was divided into three battalions, each composed of four companies. A company was divided into "squadrons" for easy maneuvering on the field. The cavalry regiment was much more expensive to sustain while in service due to the amount of equipment carried by each cavalryman (carbine, saber, pistol, belt set, and equipment for the soldier's mount) and the requirement for horses and their care.

Mathew Brady photo from the Edward Eaton Collection

The supply and maintenance of the army was a huge task. Wagon trains carrying food, ammunition, and equipment stretched for miles behind the marching columns and camped near the troops as seen in this photo of a Union army encampment in Virginia.

Both armies also had a compliment of quartermaster, engineer, and signal units as well as supply wagons organized as "trains". An army on the march was usually followed by miles and miles of wagons loaded with the equipment of war including food, ammunition, and medical supplies. At the top of the organizational list was the Army Headquarters. The commanding general required a personal staff to dictate orders and keep records of army movement. There were also clerks and assistants. The commanders of armies also had the privilege of a headquarters cook. Every army headquarters usually had a large compliment of staff officers, couriers, and a headquarters guard, which included an infantry battalion and a cavalry escort.

11

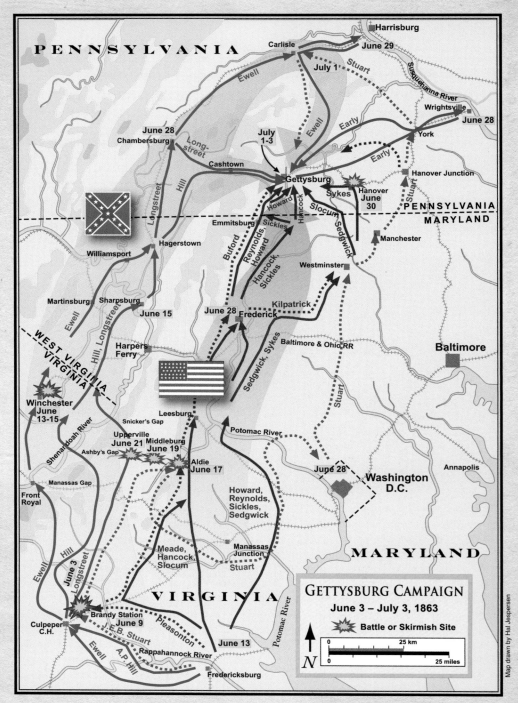

PENNSYLVANIA

Harrisburg

Carlisle — June 29

Stuart — July 1

Ewell

Susquehanna River

Wrightsville

June 28

York

June 28
Chambersburg

Long-
street

Ewell

Early

Hanover Junction

July
1-3

Cashtown

Longstreet

Hill

Gettysburg

Sykes

Howard
Hancock

Slocum

Hanover
June 30

Early

Stuart

PENNSYLVANIA

July
1-3

MARYLAND

Emmitsburg

Sickles

Sedgwick

Hagerstown

Buford

Reynolds,
Howard

Manchester

Williamsport

Hancock,
Sickles

Westminster

Martinsburg

Sharpsburg

June 15

June 28
Frederick

Kilpatrick

Ewell

Hill, Longstreet

Harpers
Ferry

Baltimore & Ohio RR

Baltimore

WEST VIRGINIA
VIRGINIA

Sedgwick, Sykes

Stuart

Winchester
June
13-15

Shenandoah River

Leesburg

Snicker's Gap

Potomac River

Annapolis

Upperville
June 21

Middleburg
June 19

June 28

Washington
D.C.

Ashby's Gap

Aldie
June 17

Manassas Gap

Howard,
Reynolds,
Sickles,
Sedgwick

Front
Royal

MARYLAND

Hill

Ewell

Longstreet

Meade,
Hancock,
Slocum

Manassas
Junction

Stuart

June 3

VIRGINIA

Brandy Station
June 9

Pleasonton

Potomac River

GETTYSBURG CAMPAIGN

June 3 – July 3, 1863

💥 Battle or Skirmish Site

Culpeper
C.H.

J.E.B. Stuart

Ewell

A.P. Hill

Rappahannock River

June 13

0   25 km

N   0   25 miles

Fredericksburg

Map drawn by Hal Jespersen

The Plan of Campaign

Lee's plan of campaign was undoubtedly similar to that of his invasion which ended in the battle of Antietam in September 1862. He then called attention to the need of destroying the bridge over the Susquehanna River at Harrisburg and of disabling the Pennsylvania Railroad in order to sever communication with the west. "After that," he added, "I can turn my attention to Philadelphia, Baltimore, or Washington as may seem best for our interest."

Kurz & Allison Chromolithograph 1889

Confederate General "Stonewall" Jackson was mortally wounded at Chancellorsville in May 1863.

Lee had suffered an irreparable loss at Chancellorsville when "Stonewall" Jackson was mortally wounded. Now reorganized into three infantry corps under Longstreet, A. P. Hill, and R. S. Ewell, and a cavalry division under J. E. B. Stuart, a changed Army of Northern Virginia faced the great test that lay ahead. "Stonewall" Jackson, the right hand of Lee, and in the words of the latter "the finest executive officer the sun ever shone on," was no longer present to lead his corps in battle.

The long lines of gray started moving on June 3 from Fredericksburg, Va., first northwestward across the Blue Ridge, then northward in the Shenandoah Valley. On June 9, one of the greatest cavalry engagements of the war occurred at Brandy Station. Union horsemen, for the first time, held Stuart's men on even terms. The Confederates then continued their march northward, with the right flank constantly protected by Stuart's cavalry, which occupied the passes of the Blue Ridge. Stuart was ordered to hold these mountain gaps until the advance into Pennsylvania had drawn the Union Army north of the Potomac. On June 28, Hill and Longstreet reached Chambersburg, 16 miles north of the Pennsylvania boundary. Rodes' division of Ewell's corps reached Carlisle on June 27. Early's command of 8,000 men had passed through Gettysburg on June 26 and on the 28th had reached York. Early planned to take possession of the bridge over the Susquehanna at Columbia, and to move on Harrisburg from the east. Lee's converging movement on Harrisburg seemed to be on the eve of success.

An unforeseen shift of events between June 25 and 28, however, threatened to deprive Lee of every advantage he had thus far gained in his daring march up the Shenandoah and Cumberland Valleys. The cavalry engagement between Stuart and Pleasonton at Brandy Station convinced Gen. Joseph Hooker, then in command of the Union Army, that

General Hooker

the Confederate Army was moving northward. President Lincoln and General in Chief Halleck, informed of this movement, ordered Hooker to proceed northward and to keep his command between the Confederate Army and Washington. When he was refused permission to abandon Harpers Ferry, and to add the garrison of 10,000 men to his army, Hooker asked to be relieved of command. Gen. George G. Meade received orders to assume command of the army at Frederick, Md., on June 28, and he at once continued the march northward.

General Stuart, in command of the Confederate cavalry, had obtained conditional approval from Lee to operate against the rear of the Union Army as it marched northward and then to join Lee north of the Potomac. As he passed between Hooker's army and Washington, the unexpected speed of the Union Army forced Stuart into detours and delays, so that on June 28 he was in eastern Maryland, wholly out of touch with the Confederate force. The eyes and ears of Lee were thus closed at a time when their efficient functioning was badly needed.

In this state of affairs, a Confederate agent reported to Lee at Chambersburg, Pa., on the night of June 28, that the Union forces had crossed the Potomac and were in the vicinity of Frederick. With the entire Union Army close at hand and with many miles between

him and his base, Lee decided to abandon his original plan and to concentrate for battle. He moved his army at once across the mountains to Cashtown, 8 miles from Gettysburg. Here, near Cashtown, he planned to establish his battle position. Rodes, then at Carlisle, and Early, at York, were at once ordered to this point.

Edwin Forbes drawing / Library of Congress

14

General Stuart, commander of the Confederate cavalry, operated against the rear of the Union Army as it marched northward and planned to join Lee north of the Potomac.

15

Order of Battle
ARMY OF THE POTOMAC
General George G. Meade
93,000 Men

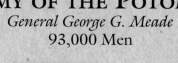

I CORPS *Reynolds*
First Division: Wadsworth
Brigades: Meredith, Cutler
Second Division: Robinson
Brigades: Paul, Baxter
Third Division: Doubleday
Brigades: Rowley, Stone, Stannard

V CORPS *Sykes*
First Division: Barnes
Brigades: Tilton, Sweitzer, Vincent
Second Division: Ayres
Brigades: Day, Burbank, Weed
Third Division: Crawford
Brigades: McCandless, Fisher

XII CORPS *Slocum*
First Division: Williams
Brigades: McDougall,
Lockwood, Ruger
Second Division: Geary
Brigades: Candy, Cobham,
Greene

II CORPS *Hancock*
First Division: Caldwell
Brigades: Cross, Kelly, Zook, Brooke
Second Division: Gibbon
Brigades: Harrow, Webb, Hall
Third Division: Hays
Brigades: Carroll, Smyth, Willard

VI CORPS *Sedgwick*
First Division: Wright
Brigades: Torbert, Bartlett, Russell
Second Division: Howe
Brigades: Grant, Neill
Third Division: Newton
Brigades: Shaler, Eustis, Wheaton

CAVALRY CORPS
Pleasonton
First Division: Buford
Brigades: Gamble, Devin,
Merritt
Second Division: Gregg
Brigades: McIntosh, Huey,
Gregg
Third Division: Kilpatrick
Brigades: Farnsworth, Custer

III CORPS *Sickles*
First Division: Birney
Brigades: Graham, Ward, De
Trobriand
Second Division: Humphreys
Brigades: Carr, Brewster, Burling

XI CORPS *Howard*
First Division: Barlow
Brigades: Von Gilsa, Ames
Second Division: Von Steinwehr
Brigades: Coster, Smith
Third Division: Schurz
Brigades: Schimmelfennig,
Krzyzanowski

Order of Battle

Army of Northern Virginia
General Robert E. Lee
75,000 Men

First Corps *Longstreet*

McLaws' Division
Kershaw's Brigade
Semmes' Brigade
Barksdale's Brigade
Wofford's Brigade

Pickett's Division
Garnett's Brigade
Kemper's Brigade
Armistead's Brigade

Hood's Division
Law's Brigade
Robertson's Brigade
G.T. Anderson's Brigade
Benning's Brigade

Second Corps *Ewell*

Early's Division
Hay's Brigade
Smith's Brigade
Gordon's Brigade
Avery's Brigade

Johnson's Division
Steuart's Brigade
Walker's Brigade
Williams' Brigade
Jones' Brigade

Rodes' Division
Daniel's Brigade
Iverson's Brigade
Doles' Brigade
Ramseur's Brigade
O'Neal's Brigade

Third Corps *Hill*

R.H. Anderson's Division
Wilcox's Brigade
Wright's Brigade
Mahone's Brigade
Lang's Brigade
Posey's Brigade

Heth's Division
Pettigrew's Brigade
Brockenbrough's Brigade
Archer's Brigade
Davis' Brigade

Pender's Division
Perrin's Brigade
Lane's Brigade
Thomas' Brigade
Scales' Brigade

Cavalry
Stuart

Hampton's Brigade
F. Lee's Brigade

Robertson's Brigade
Jenkins' Brigade

Jones' Brigade
W.H.F. Lee's Brigade

John Buford dismounted his cavalry along McPherson's Ridge and held back the Confederate advance on Gettysburg on the morning of July 1.

John Buford

The First Day, Wednesday July 1, 1863

THE TWO ARMIES CONVERGE ON GETTYSBURG. The men of Heth's division, leading the Confederate advance across the mountain, reached Cashtown on June 29. Pettigrew's brigade was sent on to Gettysburg the following day to obtain supplies, but upon reaching the ridge a mile west of the town, they observed a column of Union cavalry approaching. Not having orders to bring on an engagement, Pettigrew withdrew to Cashtown.

In the intervening 2 days since he had assumed command of the Union forces, General Meade had moved his troops northward and instructed his engineers to survey a defensive battle position at Pipe Creek, near Taneytown, in northern Maryland. Buford's cavalry, which had effectively shadowed Lee's advance from the mountaintops of the Blue Ridge, was ordered to make a reconnaissance in the Gettysburg area. It was these troops that Pettigrew's men saw posted on the roads leading into the town, Neither Lee nor Meade yet foresaw Gettysburg as a field of battle, Each expected to take a strong defensive position and force his adversary to attack.

A. P. Hill, in the absence of Lee, who was still beyond the mountains, now took the initiative. At daybreak of July 1, he ordered the brigades of Archer and Davis, of Heth's

Elements of the two armies collide west of Gettysburg during the early morning hours.
The fighting escalates through the day as more Union and Confederate troops reach the field.
By 4 pm the defending Union troops are defeated and retreat through Gettysburg where
many are captured. The remnants of the Union force rally upon Cemetery and Culp's hills.

division, to advance along the Chambersburg Road to Gettysburg for the purpose of testing the strength of the Union forces. As these troops reached Marsh Creek, 4 miles from Gettysburg, they were fired upon by Union cavalry pickets who hurriedly retired to inform their commander of the enemy's approach. In the meantime, Buford's division of cavalry had moved from their camp just southwest of Gettysburg to McPherson Ridge, a mile west of the town, Buford prepared to hold out against heavy odds until aid arrived. Thus subordinate field commanders had chosen the ground for battle.

Union General John Reynolds, arriving with his 1st Corps to relieve Buford's men, was close to the front lines and killed instantly early in the action in Herbst's Woods.

It was 8 a. m., July 1, when the two brigades of Archer and Davis, the former to the right and the latter to the left of the Chambersburg Road, deployed on Herr Ridge. Supported by Pegram's artillery, they charged down the long slope and across Willoughby Run against Buford's men. The cavalry had an advantage in their rapid-fire, breech-loading carbines. Dismounted, and fighting as infantrymen, they held their ground against the spirited attacks of Heth's superior numbers. At 10 o'clock timely aid arrived as troops from Gen. John F. Reynolds' First Infantry corps began streaming over Seminary Ridge from the south and relieved Buford's exhausted fighters. Calef's battery, one of whose guns had fired the first Union cannon shot at Gettysburg, was replaced by Hall's Maine artillery. But, in a few moments, Union joy at receiving aid was offset by tragedy. Reynolds, close to the front lines, was killed instantly by a sharpshooter's bullet.

The struggle increased in scope as more forces reached the field. When Archer's Confederates renewed the attack across Willoughby Run, Union troops of Meredith's Iron Brigade, arriving opportunely, struck the flank of the Confederates, routing them and capturing close to 100 men, including General Archer. Relieved from the threat south of the Chambersburg Pike, the 14th Brooklyn and 95th New York regiments shifted to the north of the Pike where the Confederates were overwhelming the Union defenders. With renewed effort, these troops, joined by Dawes' 6th Wisconsin, drove the Confederates steadily back, capturing 200 Mississippians in a railroad car. The Confederates then withdrew beyond striking distance, There was a lull in the fighting during the noon hour. The first encounter had given Union men confidence. They had held their ground against superior numbers and had captured Archer, a brigadier general, the first Confederate general officer taken since Lee assumed command.

THE BATTLE OF OAK RIDGE. While the initial test of strength was being determined west of Gettysburg by advance units, the main bulk of the two armies was pounding over the roads from the north and south, converging upon the ground chosen by Buford. Rodes' Confederates, hurrying southward from Carlisle to meet Lee at Cashtown, received orders at Biglerville to march to Gettysburg. Confederate General Early, returning from York with Cashtown as his objective, learned at Heidlersburg of the action at Gettysburg and was ordered to approach by way of the Harrisburg Road.

Employing the wooded ridge as a screen from Union cavalry north of Gettysburg, Rodes brought his guns into position on Oak Ridge about 1 o'clock and opened fire on the flank of Gen. Abner Doubleday, Reynolds' successor, on McPherson Ridge. The Union commander shifted his lines northeastward to Oak Ridge and the Mummasburg Road to meet the new attack. Rodes' Confederates struck the Union positions at the stone wall on the ridge, but the attack was not well coordinated and resulted in failure. Iverson's brigade was nearly annihilated as it made a left wheel to strike from the west. In the meantime, more Union troops had arrived on the field by way of the Taneytown Road. Two divisions of Howard's Eleventh corps were now taking position in the plain north of the town, intending to make contact with Doubleday's troops on Oak Ridge.

Doles' Confederate brigade charged across the plain and was able to force Howard's troops back temporarily, but it was the opportune approach of Early's division from the northeast on the Harrisburg Road which rendered the Union position north of Gettysburg indefensible. Arriving in the early afternoon as the Union men were

Looking east toward McPherson's Ridge and Herbst's Woods. General Reynolds was killed in this edge of the woods. The cupola of the Lutheran Theological Seminary appears in the background on Seminary Ridge.

Cupola of the Lutheran Theological Seminary

Assault of Brockenbrough's Confederate brigade (from Heth's Division) upon the barn of the McPherson farm in the afternoon of July 1st.

McPherson barn today.

Battles & Leaders of the Civil War

establishing their position. Early struck with tremendous force, first with his artillery and then with his infantry, against General Barlow. Soon he had shattered the entire Union force. The remnants broke and turned southward through Gettysburg in the direction of Cemetery Hill. In this headlong and disorganized flight General Schimmelfenning was lost from his command, and, finding refuge in a shed, he lay 2 days concealed within the Confederate lines. In the path of Early's onslaught lay the youthful Brigadier Barlow severely wounded, and the gallant Lieutenant Bayard Wilkeson, whose battery had long stood against overwhelming odds, mortally wounded.

The Union men on Oak Ridge, faced with the danger that Doles would cut off their line of retreat, gave way and retired through Gettysburg to Cemetery Hill. The withdrawal of the Union troops from the north and northwest left the Union position on McPherson Ridge untenable. Early in the afternoon, when Rodes opened fire from Oak Hill, Heth had renewed

19 year-old Lieutenant Bayard Wilkeson directs his Union battery at the height of the battle on July 1. Severely wounded, he refused to leave his post until loss of blood forced him to turn over his command. He died that evening in the nearby Alms House.

his thrust along the Chambersburg Pike. His troops were soon relieved and Pender's division, striking north and south of the road, broke the Union line. The Union troops first withdrew to Seminary Ridge, then across the fields to Cemetery Hill. Here was advantageous ground which had been selected as a rallying point if the men were forced to relinquish the ground west and north of the town. Thus, by 5 o'clock, the remnants of the Union forces (some 6,000 out of the 18,000 engaged in the first day's struggle) were on the hills south of Gettysburg.

Ewell was now in possession of the town, and he extended his line from the streets eastward to Rock Creek. Studiously observing the hills in his front, he came within range of a Union sharpshooter, for suddenly he heard the thud of a minie ball. Calmly riding on, he remarked to General Gordon at his side, "You see how much better fixed for a fight I am than you are. It don't hurt at all to be shot in a wooden leg."

A momentous decision now had to be made. Lee had reached the field at 3 p. m., and had witnessed the retreat of the disorganized Union troops through the streets of Gettysburg. Through his glasses he had watched their attempt to reestablish their lines on Cemetery Hill. Sensing his advantage and a great opportunity, he sent orders to Ewell by a staff officer to "press those people" and secure the hill (Cemetery Hill) if possible. However, two of Ewell's divisions, those of Rodes and Early, had been heavily engaged throughout the afternoon and were not well in hand. Johnson's division could not reach the field until late in the evening, and the reconnaissance service of Stuart's cavalry was not yet available. General Ewell, uninformed of the Union strength in the rear of the hills south of Gettysburg, decided to await the arrival of Johnson's division. Cemetery Hill was not attacked, and Johnson, coming up late in the evening, stopped at the base of Culp's Hill. Thus passed Lee's opportunity of July 1.

When the Union troops retreated from the battleground north and west of the town on the evening of July 1, they hastily occupied defense positions on Cemetery Hill, Culp's Hill, and a part of Cemetery Ridge. Upon the arrival of Slocum by the Baltimore Pike and Sickles by way of the Emmitsburg Road, the Union right flank at Culp's Hill and Spangler's Spring and the important position at Little Round Top on the left were consolidated. Thus was developed a strong defensive battle line in the shape of a fish hook, about 3 miles long, with the advantage of high ground and of interior lines.

This photo is from McPherson's Ridge, on the Chambersburg Pike. Herr's Ridge, from where Confederate Artillery first opened on McPherson's Ridge, is approximately three-fourths of a mile in the distance. This view was taken facing west in the 1880s by William Tipton.

Painting by Peter F. Rothermel in The State Museum of Pennsylvania, Pennsylvania Historical and Museum Commission

Opposite, in a semi-circle about 6 miles long, extending down Seminary Ridge and into the streets of Gettysburg, stood the Confederates who, during the night, had closed in from the north and west.

The greater part of the citizenry of Gettysburg, despite the prospect of battle in their own yards, chose to remain in their homes. Both army commanders respected noncombatant rights to a marked degree. Thus, in contrast with the fields of carnage all about the village, life and property of the civilian population remained unharmed, while the doors of churches, schools, and homes were opened for the care of the wounded.

Peter Rothermel's painting captured the chaos of the July 1 battlefield. General Reynolds is being carried from the field in the foreground.

General Meade, at Taneytown, had learned early in the afternoon of July 1 that a battle was developing and that Reynolds had been killed, A large part of his army was within 5 miles of Gettysburg. Meade then sent General Hancock to study and report on the situation. Hancock reached the field just as the Union troops were falling back to Cemetery Hill. He helped to rally the troops and left at 6 o'clock to report to Meade that in his opinion the battle should be fought at Gettysburg. Meade acted on this recommendation and immediately ordered the concentration of the Union forces at that place. Meade himself arrived near midnight on July 1.

27

Library of Congress

The Mary Thompson house on the Chambersburg Pike was used by General Lee as his headquarters. It has been carefully restored to its wartime appearance.

Library of Congress

Union General George Meade used Lydia Leister's home on the Taneytown Road as his headquarters. The Leister house has been restored and is maintained by the National Park Service.

NPS / Gettysburg NMP

Gettysburg in 1863 as seen from Seminary Ridge.

On July 1, John Burns, a 69 year old citizen of Gettysburg, took up his musket and powder horn and walked out to the scene of the morning fighting. He managed to secure a more modern rifled musket and some ammunition. Burns made his way up to the 150th Pennsylvania Infantry and asked permission to fight with them. They sent him over to the 7th Wisconsin in Herbst's Woods, where Burns participated in the afternoon fighting and was wounded. After the battle he became a national hero.

Federal dead on the field of battle.

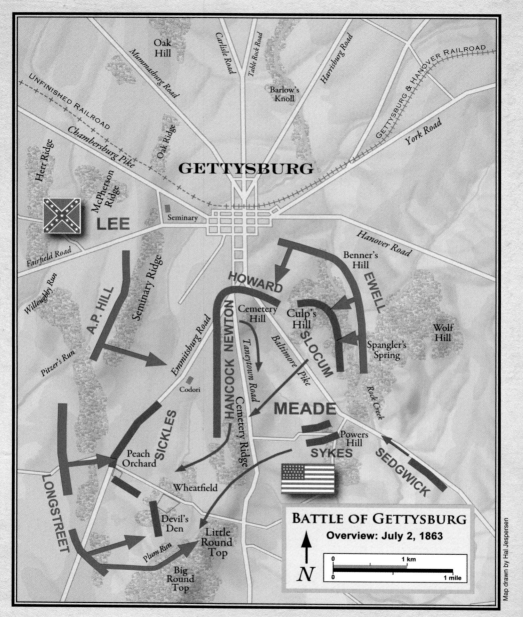

The main strength of both armies has arrived on the field by the morning hours. General Lee launches attacks against the Union left and right in an attempt to dislodge Meade's army from its strong position. Longstreet's assault upon the Union left makes good progress but is eventually checked by Federal reinforcements from the center and right. On the Union right, Ewell's Confederate troops are able to seize a foothold of Culp's Hill; elsewhere they are repulsed.

The Lutheran Seminary sits on top of Seminary Ridge. The cupola was used by both Union and Confederate signalmen during the battle because of its panoramic view of the country.

The Second Day, *Thursday July 2, 1863*

PRELIMINARY MOVEMENTS AND PLANS. The small college town of Gettysburg, with 2,400 residents at the time of the battle, lay in the heart of a fertile country, surrounded by broad acres of crops and pastures. Substantial houses of industrious Pennsylvania farmers dotted the countryside. South of the town and hardly more than a musket shot from the houses on its outer edge, Cemetery Hill rose somewhat abruptly from the lower ground. Extending southward from the hill for nearly 2 miles was a long roll of land called Cemetery Ridge. At its southern extremity a sharp incline terminated in the wooded crest of Little Round Top and a half mile beyond was the sugar-loaf peak of Big Round Top, the highest point in the vicinity of Gettysburg. Paralleling Cemetery Ridge, at an average distance of two-thirds of a mile to the west, lay Seminary Ridge, which derived its name from the Lutheran Seminary that stood upon its crest a half mile west of Gettysburg. In 1863, 10 roads radiated from Gettysburg, the one leading to Emmitsburg extending diagonally across the valley between Seminary and Cemetery Ridges.

By noon of July 2, the powerful forces of Meade and Lee were at hand, and battle on a tremendous scale was imminent. That part of the Union line extending from Cemetery Hill to Little Round Top was strongly held. Late in the forenoon, Sickles, commanding the Third Corps which lay north of Little Round Top, sent Berdan's sharpshooters and some of the men of the 3rd Maine Regiment forward from the Emmitsburg Road to

While Confederate artillery fire from Seminary Ridge, marked by the tree line in the distance, Union 3rd Corps commander, General Daniel Sickles, gallops forward toward his infantry positioned near the Peach Orchard.

Pitzer's Woods, a half mile to the west. As they reached the woods, a strong Confederate force fired upon them, and they hurriedly retired to inform their commander. To Sickles, the extension of the Confederate line southward meant that his left flank was endangered. He at once began moving forward to the advantageous high ground at the Peach Orchard, and by 3:30 p. m. his battle front extended from Devil's Den northwestward to the Orchard and northward on the Emmitsburg Road. In this forward movement, the strong position on the crest of Little Round Top was left unoccupied. This was the situation when Meade finally turned his attention from his right flank at Culp's Hill and Spangler's Spring—the cause of his great concern throughout the forenoon—to review Sickles' line.

Lee planned to attack, despite the advice of Longstreet who continually urged defensive battle. On July 2, Longstreet recommended that Lee swing around the Union left at Little Round Top, select a good position, and await attack. Lee observed that while the Union position was strong if held in sufficient numbers to utilize the advantage of interior lines, it presented grave difficulties to a weak defending force. A secure lodgment on the shank of the hook might render it possible to sever the Union Army and to deal

Trostle farmhouse. Here the 9th Massachusetts battery, taking position in the yard, lost 80 out of 88 horses during the battle of July 2. Gardner photograph.

Trostle farmhouse today.

with each unit separately. Not all of Meade's force had reached the field, and Lee thought he had the opportunity of destroying his adversary in the process of concentration. He resolved to send Longstreet against the Federal left flank which he believed was then on lower Cemetery Ridge, while Ewell was to storm Cemetery Hill and Culp's Hill.

LONGSTREET ATTACKS ON THE RIGHT. In the execution of this plan, Longstreet was ordered to take position across the Emmitsburg Road and to attack what was thought to be the left flank of the Union line on Cemetery Ridge. From his encampment on the Chambersburg Road, 3 miles west of Gettysburg, he started toward his objective, using Herr Ridge to conceal the movement from Union signalmen on Little Round Top. After marching to Black Horse Tavern on the Fairfield Road, he realized that his troops were in sight of the signal unit and at once began retracing his course. Employing the trees on Seminary Ridge as a screen, he marched southward again in Willoughby Run Valley, arriving in position on the Emmitsburg Road about 3:30 p.m. Immediately in front, and only 700 yards away, Longstreet saw Sickles' batteries lined up in the Peach Orchard and on the Emmitsburg Road. Col. E. P. Alexander, commanding Longstreet's artillery battalions, opened with full force against the Union guns. A moment later, Law's Alabama brigade stepped off, with Robertson's Texans on the left. They advanced east, then swung toward the north, with Devil's Den and the Round Tops in their path.

Major General
Gouverneur K. Warren

WARREN SAVES LITTLE ROUND TOP. Gen. G. K. Warren, Meade's Chief of Engineers, after reviewing Sickles' line with Meade, rode to the crest of Little Round Top and found the hill, "the key to the Union position," unoccupied except by a signal station. Warren was informed by the signalmen that they believed Confederate troops lay concealed on the wooded ridge a mile to the west. Smith's New York battery, emplaced at Devil's Den, immediately was ordered to fire a shot into these woods. The missile, crashing through the trees, caused a sudden stir of the Confederates "which by the gleam of the reflected sunlight on their bayonets, revealed their long lines outflanking the position." Warren realized Longstreet would strike first at Little Round Top and he observed, too, the difficulty of shifting Sickles' position from Devil's Den to the hill.

At this very moment, Sykes' Fifth Corps, marching from its reserve position, began streaming across Cemetery Ridge toward the front. Warren sought aid from this corps. In answer to his plea for troops, the brigades of Vincent and Weed sprinted to Little Round Top. Law's Alabama troops were starting to scale the south slope of the hill when Vincent's men rushed to the attack. Weed's brigade, following closely, drove over the crest and engaged Robertson's Texans on the west slope. The arrival of Hazlett's battery on the summit of the hill is thus described by an eyewitness: "The passage of the six guns through the roadless woods and amongst the rocks was marvelous. Under ordinary circumstances it would have been considered an impossible fear, but the eagerness of the men . . . brought them without delay to the very summit where they went

General G. K. Warren, chief engineer of the Army of the Potomac, observes Confederate forces gathering along Warfield Ridge southwest of Little Round Top, shortly before Longstreet's attack upon the Union left began. He found the hill undefended and immediately took action to have Union infantry moved to defend it. His action helped save the Union left flank.

immediately into battle." A desperate hand-to-hand struggle ensued. Weed and Hazlett were killed, and Vincent was mortally wounded—all young soldiers of great promise.

While Law and Robertson fought on Little Round Top, their comrades struggled in the fields below. The Confederate drive was taken up in turn by the brigades of Benning, Anderson, Kershaw, Semmes, Barksdale, Wofford, Wilcox, Perry, and Wright against the divisions of three Federal corps in the Wheatfield, the Peach Orchard, and along the Emmitsburg Road. Four hours of desperate fighting broke the Peach Orchard salient, an angle in the Union line which was struck from the south and the west. It left the Wheatfield strewn with dead and wounded, and the base of Little Round Top a shambles. Sickles' men had been driven back, and Longstreet was now in possession of the west slope of Big Round Top, of Devil's Den, and the Peach Orchard. Little Round Top, that commanding landmark which, in Confederate hands would have unhinged the Union line on Cemetery Ridge, still remained in Union possession.

CULP'S HILL. In the Confederate plan, Ewell on the left was directed to attack Cemetery Hill and Culp's Hill in conjunction with Longstreet's drive. At the appointed time, the guns of Latimer's battalion on Benner's Hill, east of Gettysburg, opened a well-directed fire against the Union positions on East Cemetery Hill and Culp's Hill, but the return fire soon shattered many of Latimer's batteries and forced the remnants to retire out of range. In the final moments of this action the youthful Major Latimer was mortally wounded.

35

Joshua Chamberlain

A pivotal moment on
Gettysburg's second day
came at Little Round Top
when Colonel Joshua
Lawrence Chamberlain's
20th Maine launched a
desperate charge down the
slope. Placed at the vital far
left of the Union line, they
were soon under fierce attack
by two Alabama regiments.
A desperate battle ensued,
sometimes becoming hand-
to-hand. Finally, with his
regiment's ammunition
dwindling and nearly a
third of his men killed or
wounded, Chamberlain
ordered his regiment to fix
bayonets. The 20th Maine's
bayonet charge took the
Confederates by surprise and
drove them off Little Round
Top. Chamberlain, a
Bowdoin College professor in
civilian life, later was
awarded a Medal of Honor
for this action.

This is the last Confederate force advance on Little Round Top where Union troops from General Sykes' V Corps await the assault about 7 p.m., July 2.

Little Round Top from the northwest, with Big Round Top on the right in the distance. 1863 Brady photograph

Massachusetts Commandery Military Order of the Loyal Legion and the U.S. Army Military History Institute, Carlisle, PA

Forbes made a number of mistakes in his paintings, there was no mountain peak behind Big Round Top.   Painting by Edwin Forbes / Library of Congress

Little Round Top today, from the northwest.

The view is from Steven's Knoll, near Culp's Hill (which would be to the viewer's right) looking toward Cemetery Hill, on the early evening of July 2. General Meade and his staff are in the foreground. Meade did visit 12th Corps commander, General Henry Slocum, who

had his headquarters near here, but not at this point of the battle. The officers between Meade and the headquarters flag are Major Simon F. Barstow, Captain James Starr Humphrey, Col. James Biddle, and Captain Charles E. Cadwalader.

Dead Confederate soldier at Devils' Den. He was moved to this position by the photographer.

Devils Den, situated at the left of the Union position, was captured by the Confederates in the fierce fighting of July 2. Using the natural protection among these large boulders, Confederate sharpshooters picked off Union soldiers on the slopes of Little Round Top. A popular legend after the battle claimed that dead Confederates were found here without a trace of a wound, supposedly killed by the concussion of shells exploding among the rocks.

The reconstructed sharpshooter's barricade with Little Round Top in the distance.

Generals Meade and Winfield Hancock, commander of the 2nd Corps, on the late afternoon of July 2. The artist took liberty to depict the two generals together as there is no record they met during the fighting that day. Hancock played a critical role in helping to defeat the Confederate attacks upon the Union left and center on July 2.

Troops of Colonel Samuel Carroll's brigade sweep past the gatehouse entrance to Evergreen Cemetery to counterattack Confederate soldiers of Jubal Early's division who have reached the crest of East Cemetery Hill and threaten to capture Union artillery there.

Painting by Edwin Forbes / Library of Congress

About dusk, long after the artillery fire had ceased, Johnson's division charged the Union works on Culp's Hill. Although his right failed to make headway because of the steep incline and the strength of the Union positions, Steuart's brigade on the left had better luck. Here, on the southern slope of the hill, the Union works were thinly manned. An hour earlier, the divisions of Geary and Ruger had been called from these works to reinforce the Union center. Johnson, finding the works weakly defended, took possession of them but did nor press the attack further. Only a few hundred yards away on the Baltimore Pike lay the Union supply trains. Failure of Confederate reconnaissance here again was critically important. Thus passed another opportunity to strike a hard blow at the Union Army.

7:30 p.m., July 2, 1863: Attack of Johnston's Division, C.S.A., on the breastworks at Culp's Hill, defended by the 12th Corps brigade of Gen. George S. Greene.

Closely timed with Johnson's assault, Early's infantry started a charge toward East Cemetery Hill. Seldom if ever surpassed in its dash and desperation, Early's assault reached the crest of the hill where the defenders, as a last resort in the hand-to-hand encounter, used clubbed muskets, stones, and rammers. Long after dark, Early's Louisiana and North Carolina troops fought to hold the crest of the hill and their captured guns. But the failure of Rodes to move out of the streets of Gettysburg and attack the hill from the west enabled Hancock to shift some of his men to aid in repelling Early's attack. Faced by these Union reserves, Early's men finally gave way about 10 o'clock and sullenly retired to their lines. The Union troops stood firm.

The summit of Little Round Top provided an excellent firing position for the Parrott guns of the Battery D, 5th U.S. Artillery. Commanded by Lieutenant Charles Hazlett, the gunners had ample targets to choose from below the slope. "The country on the left and front was full of rebels," said Lieutenant Benjamin Rittenhouse, "and coming so rapidly that it seemed almost impossible to stop them." Hazlett was killed here.

Painting above and below by Edwin Forbes / Library of Congress

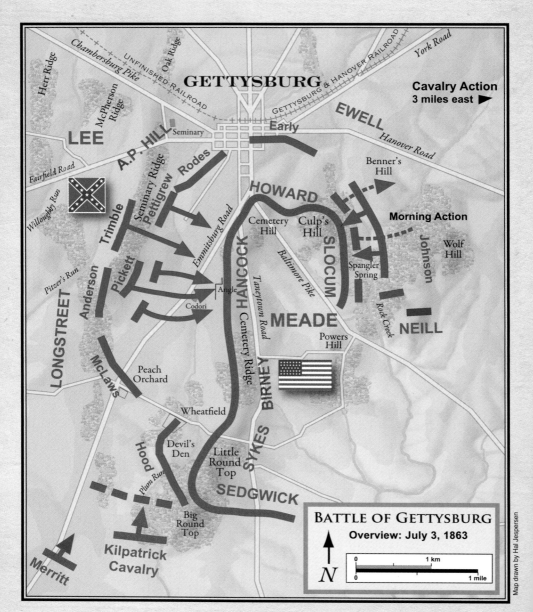

BATTLE OF GETTYSBURG

Overview: July 3, 1863

0 — 1 km
0 — 1 mile

N

Map drawn by Hal Jespersen

While Ewell renews his efforts to seize Culp's Hill, Lee turns his main attention to the Union center. Following a two-hour artillery bombardment, he sends some 12,000 Confederate infantry to try to break the Federal lines on Cemetery Ridge. Despite a courageous effort, the attack (subsequently called "Pickett's Charge") is repulsed with heavy losses. East of Gettysburg, Lee's cavalry is also checked in a large cavalry battle. Crippled by extremely heavy casualties, Lee can no longer continue the battle. On July 4 he begins to withdraw for Virginia.

48

Union General Meade with his generals at the Leister Farm House on the night of July 2

The Third Day, *Friday July 3, 1863*

CANNONADE AT DAWN: CULP'S HILL AND SPANGLER'S SPRING. Night brought an end to the bloody combat at East Cemetery Hill, but this was not the time for rest. What would Meade do? Would the Union Army remain in its established position and hold its lines at all costs? At midnight Meade sought the advice of his Council of War in the east room of his headquarters. The corps commanders—Gibbon, Williams, Sykes, Newton, Howard, Hancock, Sedgwick, and Slocum—without exception advised holding the positions established. Meade, approving, turned to the officer whose division held the Union center, and said, "Gibbon, if Lee attacks me tomorrow it will be in your front."

Despite this prediction, Meade took no unusual measures next day to fortify the center of his line. In fact, by morning he seemed convinced that the Confederate attack would continue against his left. Thus the strong forces there, three corps, were left in place. Hancock's Second Corps, holding the center, did strengthen the stone wall running along its front. And General Hunt, Chief of Artillery, brought up reserve batteries to hold in readiness for replacement of front line guns.

Meanwhile, important movements were occurring elsewhere on the field. Ruger's division and Lockwood's brigade, which had been called from their lines on the south slope of Culp's Hill the previous evening to buttress the weakened Federal forces on Cemetery Ridge, had counter-marched, under cover of darkness, to reoccupy their ground. Geary, who had misunderstood orders and had marched down the Baltimore Pike,

49

Repulse of General Johnson's Confederates near Spangler's Spring on July 3.

had also returned to his works, Ruger's men, upon reaching the Pike, learned from scouts that their entrenchments south of Culp's Hill and at Spangler's Spring had been occupied by the Confederates. Ruger, resolving upon an attack at daybreak, organized his forces along the Pike. Powerful artillery units under Muhlenberg were brought into place along the road; Rigby's Maryland battery was stationed on Power's Hill, a prominent knoll a half mile to the south; and another battery was emplaced on McAllister Hill.

Painting by Peter F. Rothermel in The State Museum of Pennsylvania, Pennsylvania Historical and Museum Commission

As dawn broke on July 3, Union guns on the Baltimore Pike opened with a heavy cannonade on Johnson's Confederates at Spangler's Spring. The heavily wooded area about the Confederate lines prevented them from bringing guns into position to return the fire. Union skirmishers began streaming across the field toward the Confederate entrenchments. The full force of Ruger's and Geary's divisions was soon committed. Throughout the forenoon the opposing lines exchanged extremely heavy fire.

About 8:30 a.m. on July 3, Jennie Wade was kneading dough for bread when a Minié ball traveled through the kitchen door of her sister's house on Baltimore Street. It pierced her left shoulder blade, went through her heart, and ended up in her corset. She was killed instantly.

General view of the rear of the Union lines, looking west on the morning of July 3rd. Troops are tramping up the Baltimore Pike, past Powers Hill to the left. Smoke rises from Cemetery Hill in the middle background and from the woods that covered Culp's Hill on the right.

It was about 10 o'clock that Ruger, believing that a flank attack might break the resistance of Johnson's men, ordered Col. Silas Colgrove to strike the Confederate left flank near the spring. The troops of the 2d Massachusetts and the 27th Indiana regiments started across the swale from the cover of the woods on the little hill south of the spring. A withering fire slowed their pace, but they charged on, only to have their ranks decimated by the Confederates in strong positions back of a stone wall. Colonel Mudge, inspiring leader of the Massachusetts regiment, fell mortally wounded. Forced to fall back, the men soon learned their efforts had not been in vain. On Ruger's and Geary's front the Confederates were now giving way and soon had retired across Rock Creek, out of striking range. By 11 o'clock, the Union troops were again in possession of their earthworks; again they could quench their thirst in the cooling waters of the spring.

LEE PLANS A FINAL THRUST. General Lee must have learned by mid-forenoon, after the long hours of struggle at Culp's Hill and Spangler's Spring, that his troops could not hold the Union works which they had occupied with so little effort the previous evening. He had seen, also, that in the tremendous battling during the preceding afternoon no important gains had been made at Little Round Top and its vicinity. Longstreet had gained the advantageous ridge at the Peach Orchard and had brought his batteries forward from Pitzer's Woods to this high ground in preparation for a follow-up attack. Wright's brigade, the last unit to move forward on July 2 in the echelon attack begun by General Law, had charged across the open fields at dusk and pierced the Union center just south of the copse of trees on Cemetery Ridge. Wright's success could not be pressed to decisive advantage as the brigades on his left had nor moved forward to his support, and he was forced to retire. Again, lack of coordination in attack was to count heavily against the Confederates.

The failure to make any pronounced headway on July 2 at Culp's Hill and Little Round Top, and the momentary success of Wright on Cemetery Ridge, doubtless led Lee to believe that Meade's flanks were strong and his center weak. A powerful drive at the center might pierce the enemy's lines and fold them back. The shattered units might then be destroyed or captured at will. Such a charge across open fields and in the face of frontal and flank fire would, Lee well understood, be a gamble seldom undertaken. Longstreet strongly voiced his objection to such a move, insisting that "no 15,000 men ever arrayed for battle can take that position."

Time now was the important element. Whatever could be done must be done quickly. Hood's and McLaws' divisions, who had fought bravely and lost heavily at Round Top and the Wheatfield, were not in condition for another severe test. Early and Johnson on the left had likewise endured long, unrelenting battle with powerful Union forces in positions of advantage. The men of Heth's and Pender's divisions had not been heavily engaged since the first day's encounter west of Gettysburg. These were the men, along with Pickett's division, whom Lee would have to count on to bear the brunt of his final great effort at Gettysburg.

LEE AND MEADE SET THE STAGE. Late in the forenoon of July 3, General Meade had completed his plan of defense. Another Confederate attack could be expected: "Where?" was still the question. General Hunt, sensing the danger, placed a formidable line of batteries in position on the crest of Cemetery Ridge and alerted others in the rear for emergency use. As a final act of preparation, Meade inspected his front at the stone wall, then rode southward to Little Round Top. There, with General Warren, he could see the long lines of massed Confederate batteries, a sure indication of attack. Meade rode back to his headquarters.

Lee, on his part, had spent the forenoon organizing his attack formations on Seminary Ridge. Having reached his decision to strike the Union center, he had ordered the movement of batteries from the rear to points of advantage. By noon, about 140 guns

Confederate infantry waiting for the end of the bombardment, before "Pickett's Charge"

were in line from the Peach Orchard northward to the Seminary buildings, many of them only 800 yards from the Union center. To Colonel Alexander fell the lot of directing the artillery fire and informing the infantry of the best opportunity to advance.

Massed to the west of Emmitsburg Road, on low ground which screened their position from the Union lines, lay Gen. George Pickett's three brigades commanded by Kemper, Armistead, and Garnett. Pickett's men had arrived the previous evening from Chambersburg, where they had guarded Lee's wagons on July 1 and 2. As a division these units had seen little fighting. Soon they would gain immortality. On Pickett's left, the attacking front was fast being organized. Joseph Pettigrew, a brigadier, was preparing to lead the division of the wounded Major General Heth, and Maj. Gen. Isaac Trimble took the command of Pender. Nearly 10,000 troops of these two divisions—including such units as the 26th North Carolina whose losses on the first day were so heavy that the dead marked their advance "with the accuracy of a line at a dress parade"—now awaited the order to attack. Many hours earlier, the Bliss farm buildings, which lay in their front, had been burned. Their objective on the ridge was in clear view. The brigades of Wilcox and Lang were to move forward on the right of Pickett in order to protect his flank as he neared the enemy position.

General Stuart, in the meantime, had been out of touch with Lee. Moving northward on the right flank of the Union Army, he became involved in a sharp engagement at Hanover, Pa., on June 30. Seeking to regain contact with Lee, he arrived at Carlisle on the evening of July 1. As he began shelling the barracks, orders arrived from Lee and he at once marched for Gettysburg, arriving north of the town the next day. Early on July 3 he was ordered to take position on the Confederate left. This movement usually has been interpreted as an integral part of Lee's assault plan. But battle reports leave Stuart's role vague, except for covering the Confederate left. Doubtless he would have exploited any significant success achieved by the infantry assault.

Except for the intermittent sniping of sharpshooters, an ominous silence prevailed over the fields. The orders had now been given; the objective had been pointed our. Men talked of casual things. Some munched on hard bread, others looked fearfully to the eastward, where, with the same mixed feelings, lay their adversary.

Far to the south, on another crucial front, General Pemberton was penning a letter to General Grant asking terms for the surrender of Vicksburg. In Richmond, the sick and anxious Jefferson Davis looked hopefully for heartening word from his great field commander at Gettysburg. The outcome of this bold venture would count heavily in the balance for the cause of the Confederacy.

ARTILLERY DUEL AT ONE O'CLOCK. At 1 p.m. two guns of Miller's Battery, posted near the Peach Orchard, opened fire in rapid succession. It was the signal for the entire line to let loose their terrific blast. Gunners rushed to their cannon, and in a few moments the massed batteries shook the countryside. Firing in volleys and in succession, the air was soon filled with smoke and heavy dust, which darkened the sky. Union gunners on Cemetery Ridge waited a few minutes until the positions of the Confederate batteries were located; then 80 guns, placed in close order, opened fire. For nearly 2 hours the duel continued, then the Union fire slackened. Hunt had ordered a partial cessation in order to cool the guns and conserve ammunition.

Colonel Alexander, in position on the Emmitsburg Road near the Peach Orchard, could observe the effectiveness of his fire on the Union lines and also keep the Confederate troops in view. To him, it appeared that Union artillery fire was weakening. His own supply of ammunition was running low. Believing this was the time to attack, Alexander sent a message to Pickett who in turn rode over to Longstreet. General Longstreet, who had persistently opposed Lee's plan of sending 15,000 men across the open ground, was now faced with a final decision. Longstreet merely nodded approval and Pickett saluted, saying, "I am going to move forward, sir." He rode back to his men and ordered the advance. With Kemper on the right, Garnett on the left, and Armistead a few yards to the rear, the division marched out in brigade front, first northeastward into the open fields, then eastward toward the Union lines. As Pickett's men came into view near the woods, Pettigrew and Trimble gave the order to advance. Sons of Virginia, Alabama,

General Pickett informs General Longstreet that his division is moving forward for its illfated attack upon the Union center on Cemetery Ridge.

57

North Carolina, Tennessee, and Mississippi, comprising the brigades of Mayo, Davis, Marshall, and Fry in front, followed closely by Lane and Lowrance, now moved out to attack. A gap between Pickett's left and Pettigrew's right would be closed as the advance progressed. The units were to converge as they approached the Union lines so that the final stage of the charge would present a solid front.

CYCLORAMA - LOOKING EAST

*Union General
Henry Hunt*

The situation is intently surveyed by Union Chief of Artillery, General Henry Hunt, (on white horse to the left of the officer with the field glasses) while General George G. Meade, commander of the Army of the Potomac, on the white horse among his staff at the edge of the field, approaches. Troops at the edge of the field beyond the officers are the Provost Guard assigned to prevent straggling. Powers Hill rises in the right background, beyond the Taneytown Road and farm fields.

CLIMAX AT GETTYSBURG. Billows of smoke lay ahead of the Union men at the stone wall, momentarily obscuring the enemy. But trained observers on Little Round Top, far to the south, could see in the rear of this curtain of smoke the waves of Confederates starting forward. Pickett finding his brigades drifting southeastward, ordered them to bear to the left, and the men turned toward the copse of trees. Kemper was now approaching on the south of the Codori buildings; Garnett and Armistead were on the north. Halted momentarily at the Emmitsburg Road to remove fence rails,

CYCLORAMA - LOOKING SOUTHEAST

Union artillery and infantry rush to fill the gap created by the Confederate breakthrough at the Angle while troops from other commands move toward the violent scene. View is Southeast, showing the several small farms on the Taneytown Road.

Pickett's troops, with Pettigrew on the left, renewed the advance. Pickett had anticipated frontal fire of artillery and infantry from the strong Union positions at the stone walls on the ridge, but now an unforeseen attack developed. Union guns as far south as Little Round Top, along with batteries on Cemetery Hill, relieved from Confederate fire at the Seminary buildings, opened on the right and left flanks. As Pickett's men drove toward the Union works at The Angle, Stannard's Vermont troops, executing a right turn movement from their position south of the copse, fired into the flank of the charging

CYCLORAMA - LOOKING SOUTH

Mounted on a black charger, General Winfield Scott Hancock directs Union troops in the defense of Cemetery Ridge, culminating at the Round Tops beyond. Soldiers crowd into the "Copse of Trees" at right center. This was the deepest penetration into the Union line by Confederate soldiers in the attack.

*Union General
Winfield Hancock*

Confederates. The advancing lines crumbled, re-formed, and again pressed ahead under terrific fire from the Union batteries.

But valor was not enough. As the attackers neared the stone wall they lost cohesion in the fury that engulfed them. All along the wall the Union infantry opened with volley after volley into the depleted ranks of Garnett and Fry. Armistead closed in, and with Lane and Lowrance joining him, made a last concerted drive. At this close range, double

CYCLORAMA - LOOKING SOUTHWEST

The "High Water Mark" has been reached. Confederate General Lewis Armistead falls mortally wounded from his horse to the right of the red battle flags as Confederate troops sweep over the stone wall. Though Armistead made the charge on foot, the artist put him on horseback so that he would stand out among the mass of combatants. The Nicholas Codori Farm buildings stand on the Emmitsburg Road in the center background.

Confederate Gen.
Lewis Armistead

65

canister and concentrated infantry fire cut wide gaps in the attacking front. Garnett was mortally wounded; Kemper was down, his lines falling away on the right and left. Armistead reached the low stone fence. In a final surge, he crossed the wall with 150 men and, with his cap on his sword, shouted "Follow me!" At the peak of the charge, he fell mortally wounded. From the ridge, Union troops rushed forward and Hall's Michigan regiments let loose a blast of musketry. The gray column was surrounded. The ride of the Confederacy had "swept to its crest, paused, and receded."

CYCLORAMA - LOOKING WEST

The view West. Pickett's, Pettigrew's and Trimble's 12,000 officers and men have crossed the mile of open fields from Seminary Ridge in the background, across the Emmitsburg Road and crashed into the Union line on Cemetery Ridge here at "The Angle" where the stone fences meet. Union troops race to recover lost ground while Confederate prisoners are pushed to the rear past an exploding artillery limber.

Two of the divisions in the charge were reduced to mere fragments. In front of the Union line, 20 fallen battle flags lay in a space of 100 yards square. Singly and in little clumps, the remnants of the gray columns that had made the magnificent charge of a few minutes earlier now sullenly retreated across the fields toward the Confederate lines. Lee, who had watched anxiously from Spangler's Woods, now rode out to meet his men. "All this has been my fault," he said to General Wilcox who had brought off his command after heavy losses. "It is I that have lost this fight, and you must help me out of it in the best

CYCLORAMA - LOOKING NORTHWEST

Confederate regiments are reduced to small groups as they charge Union troops behind by the stone wall while others, scattered in the fields beyond, flank the southern columns. General Robert E. Lee, commander of the Army of Northern Virginia, observed the last desperate moments of the charge from Seminary Ridge in far distance, where he met the survivors. The observant Union officer leaning against the tree and resting a drawn sword on his right knee is Paul Dominique Philippoteaux, the artist and creator of the Gettysburg cyclorama who signed his work with this self portrait. This view is to the northwest.

way you can." And again that night, in a moment of contemplation, he remarked to a comrade, "Too bad! too bad! Oh! too bad!"

CYCLORAMA - LOOKING NORTH

The view north on Cemetery Ridge to the Abraham Brian Farm, Ziegler's Grove, and the town of Gettysburg in the distance. Lt. William A. Arnold's Battery A, 1st Rhode Island Light Artillery fires on Pettigrew's and Trimble's troops while wounded men are evacuated from the front to the nearest medical station.

CYCLORAMA - LOOKING NORTHEAST

View looking generally Northeast toward Cemetery Hill with Culp's Hill in the right background. Amid the chaos of battle is this hospital scene, the wounded gathered by a French-influenced hay stack while surgeons operate in the open air nearby. As well as the dramatic climax of the three day battle, Philippoteaux wished to depict the suffering and somewhat primitive care of the wounded.

Union Cavalry George Custer's charge - 3 miles east of Gettysburg on July 3

CAVALRY ACTION. As the strength of Lee's mighty effort at The Angle was ebbing and the scattered remnants of the charge were seeking shelter, action of a different kind was taking place on another field not far distant. Early in the afternoon, Stuart's cavalry was making its way down the valley of Cress Run, 3 miles east of Gettysburg. The brigades of Hampton and Fitzhugh Lee, at the rear of the line of march, momentarily lost the trail and came out into open ground at the north end of Rummel's Woods, Stuart, soon learning of the mistake, attempted to bring them into line and to proceed southward. But at this point, Gen. D. M. Gregg's Union cavalry, in position along the Hanover Road a mile southeast, saw the Confederates. Gregg prepared at once to attack, and Stuart had no choice but to fight on this ground. As the two forces moved closer, dismounted men opened a brisk fire, supported by the accurate shelling of artillerists.

Then came the initial cavalry charge and countercharge. The Confederate Jenkins was forced to withdraw when his small supply of ammunition became exhausted. Hampton, Fitzhugh Lee, and Chambliss charged again and again, only to be met with the equally spirited counterattack of McIntosh. Custer's Michigan regiments assailed the front of the charging Confederate troopers, and Miller's squadron of the 3d Pennsylvania, disobeying orders to hold its position, struck opportunely on the Confederate left. The thrusts of the Union horsemen, so well coordinated, stopped the onslaught of Stuart's troopers. After 3 hours of turbulent action, the Confederates left the field and retired to the north of Gettysburg. The Union horsemen, holding their ground, had successfully cut off any prospect of Confederate cavalry aid in the rear of the Union lines on Cemetery Ridge.

Confederate prisoners photographed at Gettysburg two weeks after the battle.

"A Harvest of Death" – Stripped of shoes and other useful items by Confederates, Alexander Gardner photographed these dead Union soldiers scattered where they fell in the first day's fighting.

75

After three days of bloodletting, Lee's Army began its retreat from Gettysburg on July 4, 1863. Lee's wagon train of wounded was said to be more than seventeen miles long. This painting shows Meade's Union troops in slow pursuit of the Confederates in the following days.

End of Invasion

Lee, as he looked over the desolate field of dead and wounded and the broken remnants of his once-powerful army still ready for renewed battle, must have realized that not only was Gettysburg lost, but that eventually it might all end this way. Meade did not counterattack, as expected. The following day, July 4, the two armies lay facing each other, exhausted and torn.

Late on the afternoon of July 4, Lee began an orderly retreat. The wagon train of wounded, 17 miles in length, guarded by Imboden's cavalry, started homeward through Greenwood and Greencastle. At night, the able-bodied men marched over the Hagerstown Road by way of Monterey Pass to the Potomac. Roads had become nearly impassable from the heavy rains that day, hindering the movements of both armies. Meade, realizing that the Confederate Army was actually retreating and not retiring to the mountain passes, sent detachments of cavalry and infantry in pursuit and ordered the mountain passes west of Frederick covered. Lee, having the advantage of the more direct

Edwin Forbes painting / Library of Congress

Confederate dead.

Gardner Photograph / Library of Congress

route to the Potomac, reached the river several days ahead of his pursuers, but heavy rains had swollen the current and he could not cross. Meade arrived on the night of July 12 and prepared for a general attack. On the following night, however, the river receded and Lee crossed safely into Virginia. The Confederate Army, Meade's critics said, had been permitted to slip from the Union grasp.

Artillery Battery

Weapons and Tactics

A variety of weapons was carried at Gettysburg. Revolvers, swords, and bayonets were abundant, but the basic infantry weapon of both armies was a muzzle-loading rifle musket about 4.7 feet long, weighing approximately 9 pounds. They came in many models, but the most common and popular were the Springfield and the English-made Enfield.

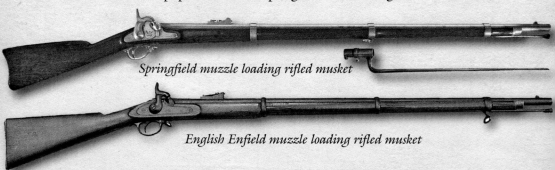

Springfield muzzle loading rifled musket

English Enfield muzzle loading rifled musket

They were hard hitting, deadly weapons, very accurate at a range of 200 yards and effective at 1,000 yards. With black powder, ignited by percussion caps, they fired "Minie Balls"— hollow-based lead slugs half an inch in diameter and an inch long. A good soldier could load and fire his rifle three times a minute, but in the confusion of battle the rate of fire was probably slower.

There were also some breech-loading small arms at Gettysburg. Union cavalrymen carried Sharps and Burnside single-shot carbines and a few infantry units carried Sharps rifles. Spencer repeating rifles were used in limited quantity by Union cavalry on July 3 and by a few Union infantry. In the total picture of the battle, the use of these efficient weapons was actually quite small.

MINIE BALLS

NEW RIFLE-MUSKET BALL
Caliber .58"

Weight Ball 500 grains.
Weight Powder 60 grains.

Those who fought at Gettysburg with rifles and carbines were supported by nearly 630 cannon—360 Union and 270 Confederate. About half of these were rifled iron pieces, all but four of the others were smoothbore bronze guns. The same types of cannon were used by both armies.

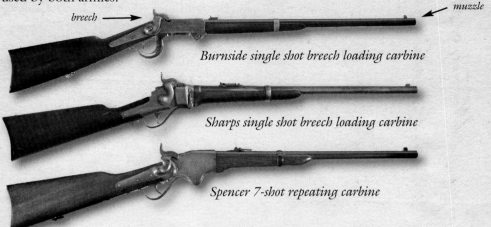

breech →
← *muzzle*

Burnside single shot breech loading carbine

Sharps single shot breech loading carbine

Spencer 7-shot repeating carbine

The majority of the bronze cannon were smoothbore howitzers and "Napoleons", the common term applied to the Model 1857 12-pounder Field Gun. They could hurl a 12-pound iron ball nearly a mile and were deadly at short ranges, particularly when firing canister. Other bronze cannon included 24 pounder howitzers and 6 pounder guns. All types are represented in the park today, coated with patina instead of being polished as they were when in use.

Most of the iron rifled pieces at Gettysburg had a 3-inch bore and fired a projectile which weighed about 10 pounds. There were two types of these— 3-inch ordnance rifles and 10 pounder Parrotts. It is easy to tell them apart for the Parrott has a reinforcing jacket around its breech, The effective range of these guns was somewhat in excess of a mile, limited in part because direct fire was used and the visibility of gunners was restricted.

This is a 3-inch ordnance rifle that fired a projectile which weighed about 10 pounds.

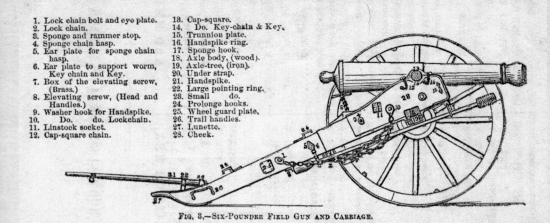

1. Lock chain bolt and eye plate.
2. Lock chain.
3. Sponge and rammer stop.
4. Sponge chain hasp.
5. Ear plate for sponge chain hasp.
6. Ear plate to support worm, Key chain and Key.
7. Box of the elevating screw, (Brass.)
8. Elevating screw, (Head and Handles.)
9. Washer hook for Handspike.
10. Do. do. Lockchain.
11. Linstock socket.
12. Cap-square chain.
13. Cap-square.
14. Do. Key-chain & Key.
15. Trunnion plate.
16. Handspike ring.
17. Sponge hook.
18. Axle body, (wood).
19. Axle-tree, (iron).
20. Under strap.
21. Handspike.
22. Large pointing ring.
23. Small do.
24. Prolonge hooks.
25. Wheel guard plate.
26. Trail handles.
27. Lunette.
28. Cheek.

FIG. 3.—SIX-POUNDER FIELD GUN AND CARRIAGE.

Two other types of rifled guns were used at Gettysburg—four bronze James guns and two Whitworth rifles. The Whitworths were unique because they were breech loading and were reported to have had exceptional range and accuracy. However, their effect at Gettysburg must have been small for one was out of action much of the time.

These artillery pieces used three types of ammunition. All cannon could fire solid projectiles or shot. They also hurled fused, hollow shells which contained black powder and sometimes held lead balls or shrapnel. Canister consisted of cans filled with iron or lead balls. These cans burst apart on firing, converting the cannon into an oversized shotgun.

This Parrott rifle has a 3-inch bore and is easy to recognize with a reinforcing jacket around its breech.

Weapons influenced tactics. At Gettysburg a regiment formed for battle, fought, and moved in a two rank line, its men shoulder to shoulder, the file closets in the rear. Since the average strength of regiments here was only about 330 officers and men, the length of a regiment's line was a little over 100 yards. Such a formation brought the regiment's slow-firing rifles together under the control of the regimental commander, enabling him to deliver a maximum of fire power at a given target. The formation's shallowness had a two-fold purpose, it permitted all ranks to fire, and it presented a target of minimum depth to the enemy's fire.

Four or five regiments were grouped into a brigade, two to five brigades formed a division. When formed for the attack, a brigade moved forward in a single or double line of regiments until it came within effective range of the enemy line. Then both parties blazed away, attempting to gain the enemy's flank if feasible, until one side or the other was forced to retire. Breastworks were erected if time permitted, but troops were handicapped in this work because entrenching tools were in short supply.

Almost all of the bronze pieces were 12 pounder "Napoleons."

Like their infantry comrades, cavalrymen also fought on foot, using their horses as means of transportation. However, mounted charges were also made in the classic fashion, particularly in the great cavalry battle on July 3.

Cavalry and infantry were closely supported by artillery. Batteries of from four to six guns occupied the crests of ridges and hills from which a field of fire could be obtained. They were usually placed in the forward lines, protected by supporting infantry regiments posted on their flanks or in their rear. Limbers containing their ammunition were nearby. Because gunners had to see their targets, artillery positions sheltered from the enemy's view were still in the future.

The procession on Baltimore Street en route to the cemetery for the dedicatory exercises, November 19.

Brady Photograph / Miller's Photographic History of the Civil War

Lincoln at Gettysburg

ESTABLISHMENT OF A BURIAL GROUND. For the residents of Gettysburg the aftermath of battle was almost as trying as the 3 days of struggle that had swirled about them. The town's 2,400 inhabitants, and the nearby country folk, bore a heavy share of the burden of caring for the 21,000 wounded and dying of both sides, who were left behind when the armies moved on. Spacious rooms in churches and schools and hundreds of homes were turned over to the care of the wounded; and kindly folk from neighboring towns came to help those of Gettysburg in ministering to the needs of the maimed and shattered men.

Adequate attention to the wounded was an immediate necessity, but fully as urgent was the need of caring for the dead. Nearly 6,000 had been killed in action, and hundreds died each day from mortal wounds. In the earlier stages of the battle, soldiers of both armies performed the tasks of burying their fallen comrades, but the struggle had reached such large proportions and the scene of battle had so shifted that fallen men had come within enemy lines. Because of the emergencies of battle, therefore, hundreds of bodies had been left unburied or only partially covered. It was evident that the limited aid which could be offered by local authorities must be supported by a well-organized plan for disinterment of the dead from the temporary burial grounds on the field and reburial in a permanent place at Gettysburg or in home cemeteries.

A few days after the battle, the Governor of the Commonwealth, Hon. Andrew Curtin, visited the battlefield to offer assistance in caring for the wounded. When official duties required his return to Harrisburg, he appointed Attorney David Wills, of Gettysburg, to act as his special agent. At the time of his visit, the Governor was especially distressed by the condition of the dead, In response to the Governor's desire that the remains be brought together in a place set aside for the purpose, Mr. Wills selected land on the

A photographer captured Lincoln (circled) on the speaker's platform during the Soldiers' National Cemetery dedication. The view shows part of the audience which was estimated at 15,000.

northern slope of Cemetery Hill and suggested that the State of Pennsylvania purchase the ground at once in order that interments could begin without delay. He proposed that contributions for the purpose of laying out and landscaping the grounds be asked from legislatures of the States whose soldiers had taken part in the battle.

Within 6 weeks, Mr. Wills had purchased 17 acres of ground on Cemetery Hill and engaged William Saunders, an eminent landscape gardener, to lay out the grounds in State lots, apportioned in size to the number of graves for the fallen of each State. Each of the Union States represented in the battle made contributions for planning and landscaping.

The reinterment of close to 3,500 Union dead was accomplished only after many months. Great care had been taken to identify the bodies on the field, and, at the time of reinterment, remains were readily identified by marked boards which had been placed at the field grave or by items found on the bodies. Even so, the names of 1,664 remained unknown, 979 of whom were without identification either by name or by State. Within a year, appropriations from the States made possible the enclosure of the cemetery with a massive stone wall and an iron fence on the Baltimore Street front, imposing gateways of iron, headstones for the graves, and a keeper's lodge. Since the original burials, the total of Civil War interments has reached 3,706. Including those of later wars, the total number now is close to 5,000.

The removal of Confederate dead from the field burial plots was not undertaken until 7 years after the battle. During the years 1870-73, upon the initiative of the Ladies Memorial Associations of Richmond, Raleigh, Savannah, and Charleston, 3,320 bodies were disinterred and sent to cemeteries in those cities for reburial, 2,935 being interred in Hollywood Cemetery, Richmond. Seventy-three bodies were reburied in home cemeteries.

Gettysburg Address

Four score and seven years ago our fathers brought forth on this continent a new nation, conceived in liberty, and dedicated to the proposition that all men are created equal.

Now we are engaged in a great civil war, testing whether that nation, or any nation, so conceived and so dedicated, can long endure. We are met on a great battle-field of that war. We have come to dedicate a portion of that field, as a final resting place for those who here gave their lives that that nation might live. It is altogether fitting and proper that we should do this.

But, in a larger sense, we can not dedicate, we can not consecrate, we can not hallow this ground. The brave men, living and dead, who struggled here, have consecrated it, far above our poor power to add or detract. The world will little note, nor long remember what we say here, but it can never forget what they did here. It is for us the living, rather, to be dedicated here to the unfinished work which they who fought here have thus far so nobly advanced. It is rather for us to be here dedicated to the great task remaining before us—that from these honored dead we take increased devotion to that cause for which they gave the last full measure of devotion—that we here highly resolve that these dead shall not have died in vain—that this nation, under God, shall have a new birth of freedom—and that government of the people, by the people, for the people, shall not perish from the earth.

November 19, 1863   *Abraham Lincoln.*

The Commonwealth of Pennsylvania incorporated the cemetery in March 1864. The cemetery "having been completed, and the care of it by Commissioners from so many states being burdensome and expensive," the Board of Commissioners, authorized by act of the General Assembly of Pennsylvania in 1868, recommended the transfer of the cemetery to the Federal Government. The Secretary of War accepted title to the cemetery for the United States Government on May 1, 1872.

DEDICATION OF THE CEMETERY. Having agreed upon a plan for the cemetery, the Commissioners believed it advisable to consecrate the grounds with appropriate ceremonies. Mr. Wills, representing the Governor of Pennsylvania, was selected to make proper arrangements for the event. With the approval of the Governors of the several States, he wrote to Hon. Edward Everett, of Massachusetts, inviting him to deliver the oration on the occasion and suggested October 23, 1863, as the date for the ceremony. Mr. Everett stated in reply that the invitation was a great compliment, but that because of the time necessary for the preparation of the oration he could not accept a date earlier than November 19. This was the date agreed upon.

Edward Everett was the outstanding orator of his day. He had been a prominent Boston minister and later a university professor. A cultured scholar, he had delivered orations on many notable occasions. In a distinguished career he became successively President of Harvard, Governor of Massachusetts, United States Senator, Minister to England, and Secretary of State.

The Gettysburg cemetery, at the time of the dedication, was not under the authority of the Federal Government. It had not occurred to those in charge, therefore, that the President of the United States might desire to attend the ceremony. When formally printed invitations were sent to a rather extended list of national figures, including the President, the acceptance from Mr. Lincoln came as a surprise. Mr. Wills was

Plan of the National Cemetery was drawn in the autumn of 1863 by the notable landscape gardener, William Saunders.

SOLDIERS' NATIONAL MONUMENT. *As a fitting memorial to the Union dead who fell at Gettysburg, the Commissioners arranged for the erection of a monument in the center of the semicircular plot of graves. A design submitted by J. G. Batterson was accepted and the services of Randolph Rogers, a distinguished American sculptor, were secured for the execution of the monument. Projecting from the four angles of the gray granite shaft are allegorical statues in white marble representing War, History, Peace, and Plenty. Surmounting the shaft is a white marble statue representing the Genius of Liberty. Known as the Soldiers' National Monument, the cornerstone was laid July 4, 1865, and the monument dedicated July 1, 1869.*

there upon instructed to request the President to take part in the program, and, on November 2, a personal invitation was addressed to him.

Throngs filled the town on the evening of November 18. The special train from Washington bearing the President arrived in Gettysburg at dusk. Mr. Lincoln was escorted to the spacious home of Mr. Wills on Center Square. Sometime later in the evening the President was serenaded, and at a late hour he retired. Ar 10 o'clock on the following morning, the appointed time for the procession to begin, Mr. Lincoln was ready. The various units of the long procession, marshaled by Ward Lamon, began moving on Baltimore Street, the President riding horse back. The elaborate order of march also included Cabinet officials, judges of the Supreme Court, high military officers, Governors, commissioners, the Vice President, the Speaker of the House of Representatives, Members of Congress, and many local groups.

Difficulty in getting the procession under way and the tardy return of Mr. Everett from his drive over the battleground accounted for a delay of an hour in the proceedings. At high noon, with thousands scurrying about for points of vantage, the ceremonies were begun with the playing of a dirge by one of the bands. As the audience stood uncovered, a prayer was offered by Rev. Thomas H. Stockton, Chaplain of the House of Representatives. "Old Hundred" was played by the Marine Band. Then Mr. Everett arose, and "stood a moment in silence, regarding the battlefield and the distant beauty of the South Mountain range." For nearly 2 hours he reviewed the funeral customs of Athens, spoke of the purposes of war, presented a detailed account of the 3-days' battle, offered tribute to those who died on the battlefield, and reminded his audience of the bonds which are common to all Americans. Upon the conclusion of his address, a hymn was sung. Then the President arose and spoke the immortal words of his Gettysburg address. A hymn was then sung and Rev. H. L. Baugher pronounced the benediction.

Four score and seven years ago our fathers brought forth on this continent, a new nation, conceived in Liberty, and dedicated to the proposition that all men are created equal.

Now we are engaged in a great civil war, testing whether that nation, or any nation so conceived and so dedicated, can long endure. We are met on a great battle field of that war. We have come to dedicate a portion of that field, as a final resting place for those who here gave their lives that that nation might live. It is altogether fitting and proper that we should do this.

But, in a larger sense, we can not dedi-

THE LINCOLN ADDRESS MEMORIAL. *The "few appropriate remarks" of Lincoln at Gettysburg came to be accepted with the passing of years not only as a fine expression of the purposes for which the war was fought, but as a masterpiece of literature. An effort to have the words of the martyred President commemorated on this battlefield culminated with the inclusion in the act approved February 12, 1895, which established Gettysburg National Military Park, of a provision for the erection of such a memorial. Pursuant to this authority, the Park Commission erected the Lincoln Address Memorial, in January 1912, near the west gate of the national cemetery.*

GENESIS OF THE GETTYSBURG ADDRESS. The theme of the Gettysburg Address was not entirely new. President Lincoln was aware of Daniel Webster's statement in 1830 that the origin of our government and the source of its power is "the people's constitution, the people's government; made for the people, made by the people, and answerable to the people." Lincoln had read Supreme Court Justice John Marshall's opinion, which states: "The government of the Union . . . is emphatically and truly a government of the people. . . Its powers are granted by them and are to be exercised directly on them, and for their benefit." In a ringing anti-slavery address in Boston in 1858, Rev. Theodore Parker, the noted minister, defined democracy as "a government of all the people, by all the people, for all the people." On a copy of this address in Lincoln's papers, this passage is encircled with pencil marks. But Lincoln did not merely repeat this theme; he transformed it into America's greatest patriotic utterance. With the Gettysburg Address, Lincoln gave meaning to the sacrifice of the dead—he gave inspiration to the living.

Rather than accept the address as a few brief notes hastily prepared on the route to Gettysburg (an assumption which has long gained much public acceptance), it should be regarded as a pronouncement of the high purpose dominant in Lincoln's thinking

89

throughout the war. Habitually cautious of words in public address, spoken or written, it is not likely that the President, on such an occasion, failed to give careful thought to the words which he would speak. After receiving the belated invitation on November 2, he yet had ample time to prepare for the occasion, and the well-known correspondent Noah Brooks stated that several days before the dedication Lincoln told him in Washington that his address would be "short, short, short" and that it was "written, but not finished."

THE FIVE AUTOGRAPH COPIES OF THE GETTYSBURG ADDRESS. Even after his arrival at Gettysburg the President continued to put finishing touches to his address. The first page of the original text was written in ink on a sheet of Executive Mansion paper. The second page, either written or revised at the Wills residence, was in pencil on a sheet of foolscap, and, according to Lincoln's secretary, Nicolay, the few words changed in pencil at the bottom of the first page were added while in Gettysburg. The second draft of the address was written in Gettysburg probably on the morning of its delivery, as it contains certain phrases that are not in the first draft but are in the reports of the address as delivered and in subsequent copies made by Lincoln. It is probable, as stated in the explanatory note accompanying the original copies of the first and second drafts in the Library of Congress, that it was the second draft which Lincoln held in his hand when he delivered the address.

Quite opposite to Lincoln's feeling, expressed soon after the delivery of the address, that it "would not scour," the President lived long enough to think better of it himself and to see it widely accepted as a master piece. Early in 1864, Mr. Everett requested him to join in presenting manuscripts of the two addresses given at Gettysburg to be bound in a volume and sold for the benefit of stricken soldiers at a Sanitary Commission Fair in New York. The draft Lincoln sent became the third autograph copy, known as the Everett-Keyes copy, and it is now in the possession of the Illinois State Historical Library.

George Bancroft requested a copy in April 1864, to be included in Autograph Leaves of Our Country's Authors. This volume was to be sold at a Soldiers' and Sailors' Sanitary Fair in Baltimore. As this fourth copy was written on both sides of the paper, it proved unusable for this purpose, and Mr. Bancroft was allowed to keep it. This autograph draft is known as the Bancroft copy, as it remained in that family for many years. It has recently been presented to the Cornell University Library. Finding that the copy written for Autograph Leaves could not be used, Mr. Lincoln wrote another, a fifth draft, which was accepted for the purpose requested. It is the only draft to which he affixed his signature. In all probability it was the last copy written by Lincoln, and because of the apparent care in its preparation it has become the standard version of the address. This draft was owned by the family of Col. Alexander Bliss, publisher of Autograph Leaves, and is known as the Bliss copy. It now hangs in the Lincoln Room of the White House, a gift of Oscar B. Cintas, former Cuban Ambassador to the United States.

Anniversary Reunions of Veterans

Over the years, the great interest of veterans and the public alike in the Gettysburg battlefield has been reflected in three outstanding anniversary celebrations. Dominant in the observance of the 25th anniversary in 1888 were the veterans themselves who returned to en camp on familiar ground. It was on this occasion that a large number of regimental monuments, erected by survivors of regiments or by states, were dedicated. Again, in 1913, on the 50th anniversary, even though the ranks were gradually thinning, the reunion brought thousands of veterans back to the battlefield. Perhaps

At the 75th anniversary of the Battle of Gettysburg, in July 1938, the Eternal Light Peace Memorial was dedicated.

the most impressive public tribute to surviving veterans occurred July 1-4, 1938, during the 75th anniversary of the battle. This was the last reunion at Gettysburg of the men who wore the blue and the gray. Although 94 years was the average age of those attending, 1,845 veterans out of a total of about 8,000 then living, returned for the encampment. It was on this occasion that the Eternal Light Peace Memorial was dedicated.

Courtesy of The Horse Soldier, Gettysburg

NPS / Gettysburg National Military Park

At the time of the seventy-fifth anniversary of the Battle of Gettysburg and of the Reunion of Confederate and Union Veterans, July 1-4, 1938, nearly 8,000 participants in the Civil War were still living. Of these, 1,845 attended the reunion. Union and Confederate veterans are here shown clasping hands across the stone wall at the Angle.

This is part of the 377' long painting of "Pickett's Charge" on display at the Visitor's Center

Photos by Bill Dow

Battle of Gettysburg Cyclorama

The "Battle of Gettysburg" cyclorama was first shown in Boston in 1884. The "Cyclorama" was a very popular form of entertainment in the late 1800s. These massive oil-on-canvas paintings were displayed in special circular auditoriums and enhanced with landscaped foregrounds sometimes featuring trees, grasses, fences and artifacts. The result is a three-dimensional effect that surrounds the viewers who stand on a central platform, literally placing them in the center of the great historic scene. French artist Paul Philippoteaux spent several weeks on the battlefield, observing details of the terrain and making hundreds of sketches. A team of assistants helped him sketch out every detail, and then began applying tons of oil paint. The phenomenal work took over a year and one-half to complete. It is shown in this book from page 58 through page 73.

Courtesy Sue Boardman

The painting first shown in Boston in 1884.

NPS / Gettysburg NMP

The painting was shown in Gettysburg from 1913 to 1962 on East Cemetery Hill.

The painting was moved to this building from 1962 to 2002 on Cemetery Hill.

The artist Paul Philippoteaux painted himself into a scene of the cyclorama..

This is a view of the McPherson farm and Herbst's Woods where Union General Reynolds was killed in July 1.

The Park

In 1895, the battlefield was established by Act of Congress as Gettysburg National Military Park. In that year, the Gettysburg Battlefield Memorial Association—founded April 30, 1864, to commemorate "the great deeds of valor . . . and the signal events which render these battlegrounds illustrious"—transferred its holdings of 600 acres to the Federal Government. In 1933, the park was transferred from the War Department to the Department of the Interior, to be administered by the National Park Service. Today, the park has some 30 miles of paved roads and an area of close to 5,000 acres. More than 1,400 monuments, tablets, and markers have been erected over the years to indicate the positions where infantry, artillery, and cavalry units fought. Hundreds of Federal and Confederate cannon are located on the field in the approximate positions of batteries during the battle. Field exhibits on the field describe important phases of the 3-day struggle.

Highlights of the Battlefield

The 24-mile route starts at the visitor center and includes the following 16 stops, the Barlow Knoll Loop, and Historic Downtown Gettysburg. The route traces the three-day battle in chronological order.

July 1, 1863

1. McPherson Ridge
The Battle of Gettysburg began about 8 am to the west beyond the McPherson barn as Union cavalry confronted Confederate infantry advancing east along Chambersburg Pike. Heavy fighting spread north and south along this ridgeline as additional forces from both sides arrived.

2. Eternal Light Peace Memorial
At 1 pm Maj. Gen. Robert E. Rodes' Confederates attacked from this hill, threatening Union forces on McPherson and Oak ridges. Seventy-five years later, over 1,800 Civil War veterans helped dedicate this memorial to "Peace Eternal in a Nation United."

3. Oak Ridge
Union soldiers here held stubbornly against Rodes' advance. By 3:30 pm, however, the entire Union line from here to McPherson Ridge had begun to crumble, finally falling back to Cemetery Hill.

When the first day ended, the Confederates held the upper hand. Lee decided to continue the offensive, pitting his 75,000-man army against Meade's Union army of 93,000.

July 2, 1863

4. North Carolina Memorial

Early in the day, the Confederate army positioned itself on high ground here along Seminary Ridge, through town, and north of Cemetery and Culp's hills. Union forces occupied Culp's and Cemetery hills, and along Cemetery Ridge south to the Round Tops. The lines of both armies formed two parallel "fishhooks".

5. Virginia Memorial

The large open field to the east is where the last Confederate assault of the battle, known as "Pickett's Charge," occurred July 3.

6. Pitzer Woods

In the afternoon of July 2, Lt. Gen. James Longstreet placed his Confederate troops along Warfield Ridge, anchoring the left of his line in these woods.

7. Warfield Ridge

Longstreet's assaults began here at 4 pm. They were directed against Union troops occupying Devil's Den, the Wheatfield, and Peach Orchard, and against Meade's undefended left flank at the Round Tops.

8. Little Round Top

Quick action by Brig. Gen. Gouverneur K. Warren, Meade's chief engineer, alerted Union officers to the Confederate threat and brought Federal reinforcements to defend this position.

9. The Wheatfield

Charge and counter-charge left this field and the nearby woods strewn with over 4.000 dead and wounded.

10. The Peach Orchard

The Union line extended from Devil's Den to here, then angled northward on Emmitsburg Road. Federal cannon bombarded Southern forces crossing the Rose Farm toward the Wheatfield until about 6:30 pm, when Confederate attacks overran this position.

11. Plum Run

While fighting raged to the south at the Wheatfield and Little Round Top, retreating Union soldiers crossed this ground on their way from the Peach Orchard to Cemetery Ridge.

12. Pennsylvania Memorial

Union artillery held the line alone here on Cemetery Ridge late in the day as Meade called for infantry from Culp's Hill and other areas to strengthen and hold the center of the Union position.

Monument at the High Water Mark, climactic moment of the battle on July 3rd

13. Spangler's Spring
About 7 pm, Confederates attacked the right flank of the Union army and occupied the lower slopes of Culp's Hill. The next morning the Confederates were driven off after seven hours of fighting.

14. East Cemetery Hill
At dusk, Union forces repelled a Confederate assault that reached the crest of this hill.

By day's end, both flanks of the Union army had been attacked and both had held, despite losing ground. In a council of war, Meade, anticipating an assault on the center of his line, determined that his army would stay and fight.

July 3, 1863

15. High Water Mark
Late in the afternoon, after a two-hour cannonade, some 7,000 Union soldiers posted around the Copse of Trees, The Angle, and the Brian Barn, repulsed the bulk of the 12,000-man "Pickett's Charge" against the Federal center. This was the climactic moment of the battle. On July 4, Lee's army began retreating.

Total casualties (killed, wounded, captured, and missing) for the three days of fighting were 23,000 for the Union army and as many as 28,000 for the Confederate army.

16. National Cemetery
This was the setting for Lincoln's Gettysburg Address, delivered at the cemetery's dedication on November 19, 1863. Use the Soldiers' National Cemetery parking area on Taneytown Road.

Historic Downtown Gettysburg Tour

A. David Wills House
Home of the prominent Gettysburg attorney who oversaw the creation of the Soldiers' National Cemetery. Abraham Lincoln finished his Gettysburg Address here the night before the cemetery dedication.

B. Gettysburg Train Station
Abraham Lincoln arrived here on November 18. This structure was also a vital part of the recovery efforts after the battle, as a depot for delivery of supplies and evacuation of the wounded.

Welcome to Gettysburg

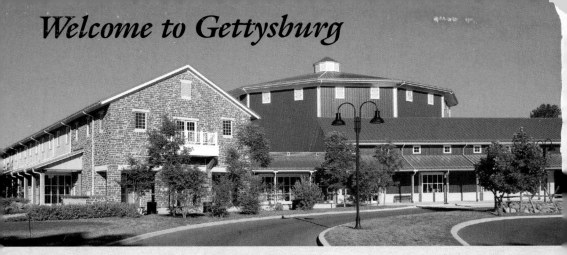

Begin your visit at the official Gettysburg National Military Park Museum and Visitor Center, located at 1195 Baltimore Pike, just south of historic downtown Gettysburg, Pennsylvania.

The Museum and Visitor Center features the exclusive Cyclorama, Film and Museum Experience. Here you will find plenty of free parking, information on how to visit the park, and what to see around Gettysburg.

At the Gettysburg National Military Park Museum and Visitor Center, you can arrange Battlefield Tours by taking a Gettysburg Licensed Battlefield Guide bus tour or a personalized Licensed Battlefield Guide car tour.

The Museum and Visitor Center at Gettysburg also sells tickets for President Dwight D. Eisenhower's home — now the Eisenhower National Historic Site — and the David Wills House in downtown Gettysburg (part of the Gettysburg National Military Park).

A variety of National Park Service Ranger-led programs are held throughout the year at Gettysburg National Military Park to help you understand the battle and its impact on the soldiers, civilians, and the nation. Check our Gettysburg events calendar for schedules.

Groups and individuals may make advance reservations for the Cyclorama, Film and Museum Experience tickets, tours with a licensed guide, and visit the Eisenhower National Historic Site by calling 1-877-874-2478 or visiting www.gettysburgfoundation.org.

The Gettysburg Foundation is a non-profit educational organization working in partnership with the National Park Service to enhance preservation and understanding of the heritage and lasting significance of Gettysburg. In addition to operating the Gettysburg National Military Park Museum and Visitor Center, the Foundation has a broad preservation mission that includes land, monument and artifact preservation and battlefield rehabilitation—all in support of the National Park Service's goals at Gettysburg.

GETTYSBURG NATIONAL MILITARY PARK

Administered by the National Park Service, United States Department of the Interior

1195 Baltimore Pike, Gettysburg, PA 17325

TICKETS: www.gettysburgfoundation.org (877) 874-2478